ICEBERGS AND THEIR VOYAGES

ICEBERGS AND THEIR VOYAGES

BY GWEN SCHULTZ

William Morrow and Company
New York 1975

BY THE SAME AUTHOR

The Blue Valentine

Inquiries should be addressed to
William Morrow and Company, Inc.,
105 Madison Ave., New York, N.Y. 10016.

Printed in the United States of America.
3 4 5

Library of Congress Cataloging in Publication Data

Schultz, Gwen M
Icebergs and their voyages.

SUMMARY: Discusses the formation, history, and location of icebergs and the possibilities for their use.
1. Icebergs—Juvenile literature. [1. Icebergs] I. Title.
GB2405.S38 551.3'42 75-9958
ISBN 0-688-22047-9
ISBN 0-688-32047-3 lib. bdg.

DEDICATED TO TEACHERS,
without whom there would be no writers,
without whom there would be no books.

ACKNOWLEDGMENTS

The author expresses appreciation to the following
for helpful information supplied:
Alaska Department of Economic Development
The Arctic Institute of North America
Danish Information Office
National Science Foundation—Office of Polar Programs
New Zealand Department of Scientific and Industrial Research—
Antarctic Division
The Rand Corporation
United States Army Cold Regions Research
and Engineering Laboratory
United States Coast Guard—International Ice Patrol,
and Public Affairs Division
United States Naval Oceanographic Office

PHOTO CREDITS

Alaska Division of Tourism, 72
American Geographical Society/William O. Field, 13, 49
Arctic Explorations/E. K. Kane, 10
Lindblad Travel, Inc., New York/George Holton, 2
NASA and United States Geological Survey Eros Data Center, 87
National Archives, 38
National Science Foundation and United States Navy, 62, 63
National Science Foundation and The Rand Corporation, 84
New Zealand Department of Scientific and Industrial Research,
Antarctic Division, 85
United States Coast Guard, 8, 12, 16, 19, 22, 25, 26, 27, 30,
32, 34, 35, 37, 39, 40, 41, 42, 43, 44, 45, 47, 48, 54,
66, 67, 68, 69, 70, 77, 93
United States Navy, 52, 56, 57, 59, 80

CONTENTS

CHAPTER ONE

FROM SNOWFLAKES TO THE SEA

Icebergs are among nature's most spectacular creations, and yet most people have never seen one. A vague air of mystery envelops them. They come into being—somewhere—in faraway, frigid waters, amid thunderous noise and splashing turbulence, which in most cases no one hears or sees. They exist only a short time and then slowly waste away just as unnoticed.

Objects of sheerest beauty, they have been called. Appearing in an endless variety of shapes, they may be dazzling white, or they may be glassy blue, green or

The perils met by wooden sailing ships in iceberg waters are portrayed in this drawing by E. K. Kane of the United States Navy in his book *Arctic Explorations,* published in 1856.

purple, tinted faintly or in darker hues. They are graceful, stately, inspiring—in calm, sunlit seas.

But they are also called frightening and dangerous, and that they are—in the night, in the fog, and in storms. Even in clear weather one is wise to stay a safe distance away from them. Most of their bulk is hidden below the water, so their underwater parts may extend out far beyond the visible top. Also, they may roll over unexpectedly, churning the waters around them. Still, people are drawn to them out of curiosity and wonderment if nothing else. Whatever their aspect, icebergs are fascinating, and we want to know more about them.

Just where do icebergs originate? What is their path and destiny? What role do they play in the environment? These questions are not idle ones, for though icebergs have always remained far from population centers, they may not all do so in the future.

The world's cold regions are being opened to resource development, commerce, tourism, and settlement. As more and more people travel and live in the realms of icebergs, interest increases in those floating masses of ice. And in ways still dimly foreseen icebergs may before long touch the lives of people even in warm regions.

To understand icebergs, one must know how they came into being. First, they should be clearly distinguished from ice floes, which are also floating ice. The two are quite different, having formed in different ways.

When the surface of the ocean (or some other sizable water body) freezes over, and then that ice layer breaks up, there are flat pieces of ice floating around. These pieces are called *ice floes,* and they make up the loose pack ice that covers oceans in polar regions. This loose pack ice may freeze together and break apart into new floes many times.

The pieces of ice called *icebergs,* on the other hand, have their origin over land, and their formation is a much slower, more involved process. They are parts of glaciers that break off, drift into the water, float about awhile, and finally melt.

Icebergs afloat today are made of snowflakes that have fallen over long ages of time. They embody snows that drifted down hundreds, or many thousands, or in some cases maybe a million years ago. The snows fell in polar regions and on cold mountains, where they melted only a little or not at all, and so collected to great depths over the years and centuries.

An iceberg sweeps a path through pack ice, which is composed of ice floes. Icebergs are fragments of glaciers. Ice floes are flat pieces of the frozen ocean surface that have broken up.

Icebergs recently broken off from Alaska's Columbia Glacier lie on the beach during low tide until the high tide returns and sets them adrift again.

As each year's snow accumulation lay on the surface, evaporation and melting caused the snowflakes slowly to lose their feathery points and become tiny grains of ice. When new snow fell on top of the old, it too turned to icy grains. So blankets of snow and ice grains mounted layer upon layer and were of such great thickness that the weight of the upper layers compressed the lower ones. (This compacting process takes place when one walks on fluffy new snow. One's weight causes the snow in the footprints to become more compact or even icy.) With time and pressure from above, the many small ice grains joined and changed to larger crystals, and eventually the deeper crystals merged into a solid mass of ice.

In this way large bodies of dense ice have formed and maintained themselves. The pressure of their weight and the pull of gravity causes them to bulge outward and spread across the land. Such moving bodies of ice formed from fallen snow are called *glaciers*. Some glaciers fill whole valleys; larger ones cover many mountains and many valleys, or broad expanses of level land, or, in the case of Greenland and Antarctica, whole land masses. Those two largest ones are called *ice sheets*.

When a glacier's outward-moving edge reaches a warm location, melting takes place and its meltwater drains away as a river. But where a spreading edge of a glacier reaches a body of water before melting, there pieces of the solid glacier break off and become icebergs. This body of water is most often the ocean, but it may also be a lake.

During the Ice Age, glaciers were more active and larger than they are today, covering as much as one third of the world's land. Then, as one would expect, there were many more icebergs in the world than there are now. At the foot of many glaciers there were huge lakes, formed mainly from meltwater draining from the ice, and icebergs were dropped into them. The vast ice sheets that covered northern North America and northern Europe advanced into the sea along fronts thousands of miles wide, as Antarctica's ice sheet still does today, and from them numerous icebergs drifted forth.

Today the areas where icebergs form are more limited. Most are in thinly populated or uninhabited parts of the world. Antarctica is the main producer of icebergs, and Greenland is second. But because Greenland has been nearer major population centers, and was within the range of sailing ships of the Vikings and other early explorers, its icebergs have been known centuries longer, and have been met oftener, than those of more isolated Antarctica.

A thousand years ago exploring seamen were encountering icebergs as they ventured across the North Atlantic. They sailed and rowed their small, light ships into hazy, mysterious northern regions, skipping from Ireland, Scotland, and Scandinavia to Iceland, Greenland, and eventually the North American continent. Later the search for the legendary Northwest Passage to the Pacific kept luring explorers into arctic waters.

Greenland is largely rimmed by mountains, which, like the sides of an oval bowl, hold a dome of ice estimated to be nearly two miles thick in its thickest parts. Snow collects continuously on the broad ice sheet, and the glacier

One of the tallest icebergs ever recorded by the United States Coast Guard is this one sighted in Melville Bay off Greenland. As measured from a helicopter, its peak is judged to be 550 feet above the water.

ice that keeps forming from it has to drain off. It presses outward through valleys, including those called *fiords* (steep-sided, ice-scoured coastal valleys), in the mountain rim. Where the ice is not confined to valleys, it flows down wider slopes to the sea. Along the curved shore of Melville Bay on Greenland's west coast, ice moves out along an almost continuous front nearly 250 miles long.

Along Greenland's west coast about 100 glaciers reach the sea. Of those, about twenty are the main iceberg makers. Some of the bergs are mammoth. They may be several city blocks long and occasionally a mile or more long. Once a berg several miles long was sighted, but it was exceptionally large. Many tower 200 to 300 feet above the water. Some over 500 feet! Greenland sends out 10,000 to 15,000 good-sized bergs each year and uncountable numbers of smaller ones. The largest are calved from the glaciers along Melville Bay and from Humboldt Glacier farther north in the northwestern part of the island.

Some smaller islands of the northern hemisphere subpolar regions also have glaciers that send out a good quantity of icebergs—Canada's Ellesmere and Devon islands just west of northern Greenland, Norway's Spitsbergen, and USSR's Novaya Zemlya in the Arctic Ocean. But their total contribution of bergs is a tiny fraction of Greenland's.

In more temperate latitudes the largest glaciers that reach the sea are those along Alaska's mountainous Pacific coast.

Columbia Glacier in Alaska west of Valdez (the end of the oil pipeline) is the largest glacier that ocean vessels can approach easily. Its ice front is about three miles wide and rises about 200 feet or more above the water. (That is about as high as a twenty-story building, as each story is about ten feet high—a good guide for estimating height.) This glacier moves seaward an average of six feet a day, casting sparkling blue fragments of ice into the water.

Glacier Bay, northwest of Juneau, is a national monument. It has several tidewater glaciers; that is, glaciers that reach the sea. Muir Glacier is the most famous. It is named for John Muir, founder of the United States National Park System, who explored that bay in 1880. The bay abounds with bergs, but most are of modest size and melt before they can leave the bay. Seals sun themselves on the low ones as swimmers do on rafts.

The Le Conte Glacier near Petersburg in southern Alaska is the most southern tidewater glacier of the northern hemisphere. Its small icebergs are sometimes seen in the main shipping channel between Petersburg and Juneau to the north.

Iceberg-dotted lakes that are not hard to reach are popular tourist centers. One such place is at Mendenhall Glacier near Juneau. There, from a comfortable glass-enclosed building, visitors can view the glacier and watch icebergs calving.

An active glacier is a powerful, impressive thing. When

part of its massive front crumbles and falls, the sight is magnificent, and one thrills at the thought of the immeasurable years and dragging slowness that led up to that brief dramatic moment.

While the brittle ice creeps downslope, the strain of its bending opens cracks and deep crevasses in it. When

Thousands of Greenland's icebergs calved from glaciers start their voyages in the spring. Some will travel long distances and last for several years. The journeys of others will be quite short.

the glacier front enters the water, its lower part is usually underwater. The movement of currents and waves wears away the base of the ice front, sometimes making it top-heavy, and gradually loosens fragments of the cracked glacier front. Piece by piece these fragments fall away.

Coming near the front of a tidewater glacier is dangerous, for there is no predicting when calving will occur. It may happen suddenly. At times icebergs break off every few minutes. However, if you are watching to see an iceberg splash into the water, you may wait endlessly without success.

The typical face of a tidewater glacier is steep and jagged with sharp points, like steeples and towers, and crack lines running down it. The first sign that an iceberg is about to be born is usually a rumbling noise and the falling of small pieces of ice down the glacier's face. Then a large section will slide down in one or more chunks accompanied by a tremendous roar. Smaller ice pieces will fall with it and after it, like an avalanche. There may be several explosions of sound as the ice breaks and falls.

Usually calving icebergs do not fall forward, but plunge straight down, or nearly so, deep into the water. The ice being lighter than water then rebounds, zooming as much as a hundred or more feet into the air with water spilling down its glistening sides like a fountain. Then the berg falls again with another roar into the turbulent water and bobs up and down until the water quiets, and finally it drifts normally among the other bergs.

Sometimes a chunk of ice detaches from the unseen, underwater part of a glacier. Without warning it explodes upward from the water into the air in a similar breathtaking display.

John Muir described the bergs he saw in Glacier Bay in these vivid words in his book *Travels in Alaska*:

> When a large mass sinks from the upper fissure portion of the wall, there is first a keen, prolonged, thundering roar, which slowly subsides into a low muttering growl, followed by numerous smaller grating, clashing sounds from the agitated bergs that dance in the waves about the newcomer as if in welcome; and these again are followed by the swash and roar of the waves that are raised and hurled up the beach. . . . But the largest and most beautiful of the bergs . . . rise from the submerged portion with a still grander commotion, springing with tremendous voice and gestures nearly to the top of the wall, tons of water streaming like hair down their sides, plunging and rising again and again before they

finally settle in perfect poise, free at last, after having formed part of the slow-moving glacier for centuries.

When sunlight glints through the freshly broken crystal ice, it shimmers with iridescent colors. The reflection of the bergs, colored or white, on a mirror surface of calm water doubles their beauty.

Some icebergs are streaked or stained with rocks and gravel picked up when they were still part of a glacier scraping across the land. They may also have huge boulders frozen in them. As they melt, this material will drop to the ocean or lake floor. Ocean-floor explorers have

Dark earth material picked up and embedded in a glacier as it scraped over the land now floats to sea in this iceberg.

found such material far from shore, including rocks bearing marks of glacial scour. They indicate how far icebergs of the past drifted.

Icebergs break loose from glaciers throughout the year, but most are released in the warm season. Then glaciers move faster; then the ocean ice parts and permits icebergs to escape into open water. In the spring ice pours forth from Greenland in superabundance. Persons who witness the great outpourings of icebergs marvel at the sight. Some of the outlet glaciers are surging glaciers, popularly known as *galloping glaciers,* which move much faster than normal—not so fast that one can see them move, but fast enough that one can notice the advance over a short period of time.

Observers of Rinks Glacier on Greenland's west coast have estimated that when it surges it dumps as much as 500,000,000 tons of ice into the ocean in a few minutes. Furthermore, it repeats that act about once every two weeks. Other outlet glaciers perform similar feats.

Icebergs that calve from glaciers at the head of a fiord may be held in the narrow inlet a long time when the ocean ice is frozen tight and they cannot float away. When that happens the trapped bergs are crowded together under increasing pressure from behind as the glacier keeps advancing and dumping off more ice. Finally when the ocean ice weakens, usually in spring, the pileup of bergs rattles loose with a violent push. There is a resounding, booming noise as they grind against each other and burst into the open. Then the ocean is filled with myriads of icebergs spreading out and drifting away.

In the Jakobshavn (Jacob's Harbor) fiord, also on Greenland's west coast, this spectacle takes place about twice a month during the summer. At the fiord's mouth is a shallow underwater ridge. The largest icebergs ground on it and serve as a cork holding back oncoming ones. They collect behind until the fiord is jammed with them. Finally the big bergs at the fiord's mouth break apart or melt or are pushed out by the uncontainable force of piled-up ice behind them, and then the fiord "shoots." It suddenly lets loose its bottled-up icebergs, and for hours the crash of colliding bergs can be heard. When the fiord has emptied, Disko Bay outside is a profusion of white. The Greenlandic name for Jakobshavn is Ilulíssat, which means "place by the icebergs."

Some icebergs form under conditions somewhat different from those described thus far. Glaciers that are solid and strong and very cold can hold together even after entering the sea, particularly if they enter it in a place that is protected from severe storms and wave action. Then they form what is called an *ice shelf,* which spreads

A giant iceberg breaks away from Greenland's Jakobshavn Glacier. It is about a mile across at the break-off line. Some bergs shown here are equal in area to several square city blocks. The bergs are drifting west through the fiord to Disko Bay in the background.

A tabular iceberg, which is a detached portion of an ice shelf. This ice island from Greenland is ¾ of a mile long and ½ of a mile wide and

out over the water. Ice shelves exist along parts of the coasts of Greenland and Ellesmere Island, and huge ones fringe Antarctica.

An ice shelf will reach as far as its strength permits. Where it reaches its limit, where it is weakened and buffeted by storms, sections of it will crack off along the outer edge. The detached pieces of ice shelf are, of course, icebergs. These bergs are tabular in shape with flat or slightly uneven tops and nearly straight, perpendicular sides. The larger ones are called *ice islands* and are different from ice floes, which are thinner and weaker. Numer-

shows 80 feet above the water. Below the surface it extends down 560 feet. Scientific camps have been built on ice islands such as this one.

ous ice islands drift in the waters around Antarctica. In the northern hemisphere there are not as many, but some float around the Arctic Ocean in irregular paths for a few years before breaking up. Seldom do they stray south of the arctic area. Scientific camps have been built on some of the arctic ice islands because they are flat and durable and also airplanes can land on them.

In such ways are icebergs born—beginning with snowflakes and tediously finding their way to the sea.

They clearly illustrate nature's hydrologic cycle, the continuous passage of water from sky to earth and back to the sky. Snow falls to form glaciers. They slide to the sea even as rain runs through rivers, and there they fragment into icebergs, which slowly liquefy. The frozen moisture they held so long is at last free again and mixes with the salty ocean. From the ocean it is evaporated and returns to the atmosphere, where it will condense into clouds and fall as droplets or snowflakes in that never-ending circular process.

HOW ICEBERGS FIT INTO THE HYDROLOGIC CYCLE: Snow falls in cold regions, feeding glaciers (which are ice formed from compressed snow). Glaciers move toward the ocean, even as rivers do, and those that reach it break up there into icebergs. Icebergs drift away and melt, and their meltwater is added to the ocean's water. Through evaporation, water at the surface changes to water vapor and is carried into the atmosphere. Air masses holding water vapor travel everywhere and, when cooled enough, produce clouds and precipitation. In cold regions snow falls, feeding glaciers, and so the cycle continues.

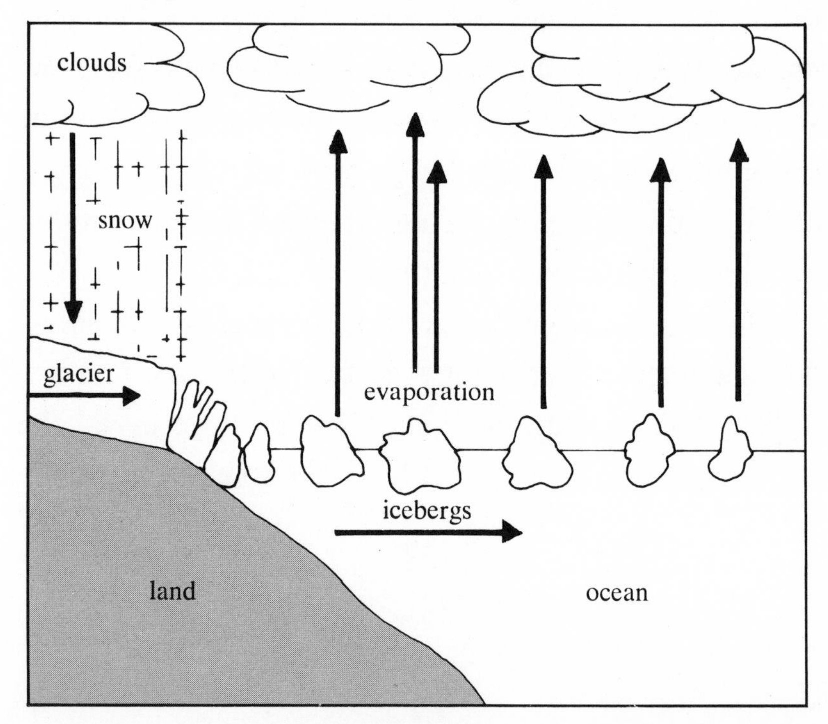

CHAPTER TWO

NORTHERN ICEBERGS

In the northern hemisphere most icebergs come from deeply crevassed, sloping glaciers that slide down to the sea and disintegrate there close to shore, so most of them are irregular in shape. The large ones are typically jagged and pinnacled, and sometimes they are even U-shaped or arched.

After the icebergs are launched, their fate depends upon which way the currents and winds carry them. Many are washed onto a shore or run onto a submarine bank, where they become stranded till they melt away.

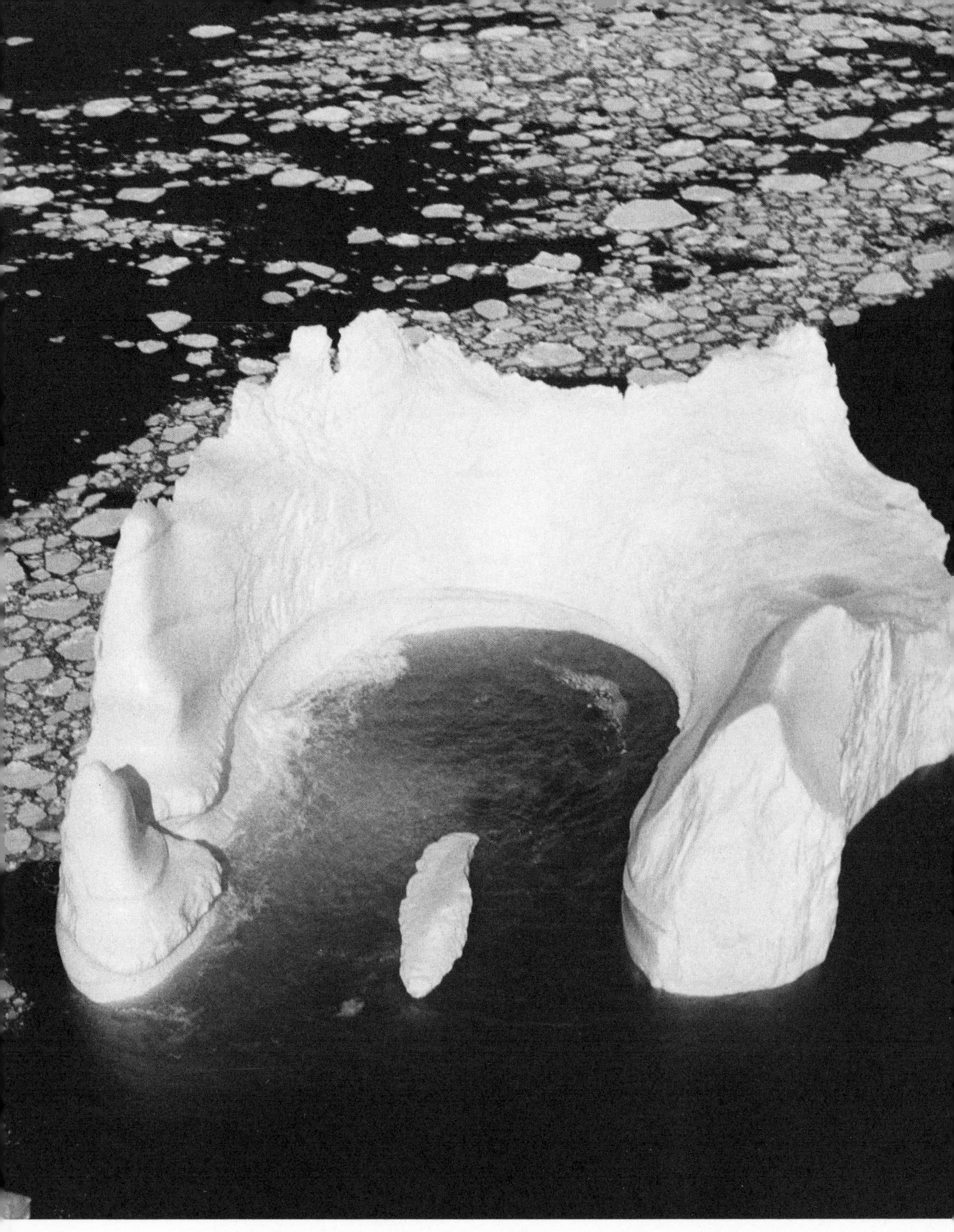

A U-shaped iceberg among ice floes on the Grand Banks.

Some batter around an island or coast for weeks before moving on, while others are carried away directly. Some drift great distances in the open water along straight or wandering routes until they too finally waste away.

The direction a berg travels is determined more by currents in the water than by wind. The reason is that 85 to 90 percent of the berg is underwater. When driven by water currents, icebergs can easily move into the wind.

To visualize how much of an iceberg is below water level, make this observation sometime. When you are having a soda or fruit drink with ice cubes in it, look closely and notice how little of the floating ice rises above the liquid. In the same way, very little of the iceberg rises above water.

The routes traveled by Greenland's icebergs, therefore, conform largely to the pattern of ocean currents in the area. Icebergs coming from the east coast of Greenland are caught in the south-moving East Greenland Current that parallels the coast and steers them around Cape Farewell, the southern tip of the island. There the East Greenland Current meets and joins the West Greenland Current, which flows northward along the west coast of the island. So, many of the icebergs from Greenland's east coast are swept that way too, northward up the west coast, traveling together with the greater number of west-

A fleet of icebergs gathered at Cape York, northwestern Greenland, ready for the drift south into the North Atlantic. Some will reach there, some not.

coast bergs. They may move north several hundred miles into Baffin Bay, northeast of Baffin Island, before swinging south. Many, however, disappear along the way in Davis Strait, the body of water between Greenland and Baffin Island. Some ground in shallow water and waste away there. Others freeze in over winter and float free again in spring.

The West Greenland Current flows north along Greenland to where the coast curves west, and there—still carrying icebergs–it curves west with the coast, and then goes south along the east coast of Baffin Island. Farther south it becomes the Labrador Current. This current, known for its coldness, transports the icebergs southward, east of Newfoundland to the Grand Banks.

The Grand Banks is a shallow part of the ocean lying just east and south of the island of Newfoundland. Its main part referred to here is somewhat larger than that island; it is about 500 miles across from east to west and extends north to south from about 48 degrees to 42 degrees north.

There the cold Labrador Current meets the warm Gulf Stream, which is coming from the Caribbean area and heading toward northern Europe. Air masses over the currents are also of contrasting temperatures, and when they merge, fog results. This spot is one of the world's leading fishing areas because of the abundance of food for fish in the cool, mixing, shallow waters, and fishing boats from many countries come here. The Grand Banks lie on the shortest ocean route between the main North Atlantic ports of Europe and North America, and this route is the busiest transoceanic shipping lane in the world.

The danger is plain. Ship traffic is heavy in these waters. There are numerous fishing boats, freighters, and passenger liners. There is fog. And there are icebergs silently

A glacier slides out from the Greenland ice sheet into Melville Bay.

drifting into the path of the ships. Collision with an iceberg is dreaded by anyone sailing through this region.

Icebergs from Greenland may survive several years and travel far. They may have come as far as 1,800 miles before reaching the Grand Banks, and many go much farther. Bergs calved from Greenland glaciers usually spend their first winter in the Melville Bay area, their second winter near Cape Dyer (southeastern Baffin Island), and they reach the Grand Banks the following spring or summer. But most that start toward the south break up and melt before getting that far. The surface temperature

of the northern edge of the Gulf Stream near the Grand Banks ranges from about 54 degrees Fahrenheit in winter to about 62 degrees in summer. After icebergs enter the Gulf Stream they seldom last more than two weeks. However, some have drifted as far south as the Azores (latitude 37° 44′ N.) and Bermuda (latitude 32° 20′ N.). In June, 1934, one was sighted in the middle of the North Atlantic at 31 degrees north, the latitude of Jacksonville, Florida, 2,000 miles south of Greenland's southern tip. It was, of course, quite small by then.

In an average year about 390 icebergs drift south of

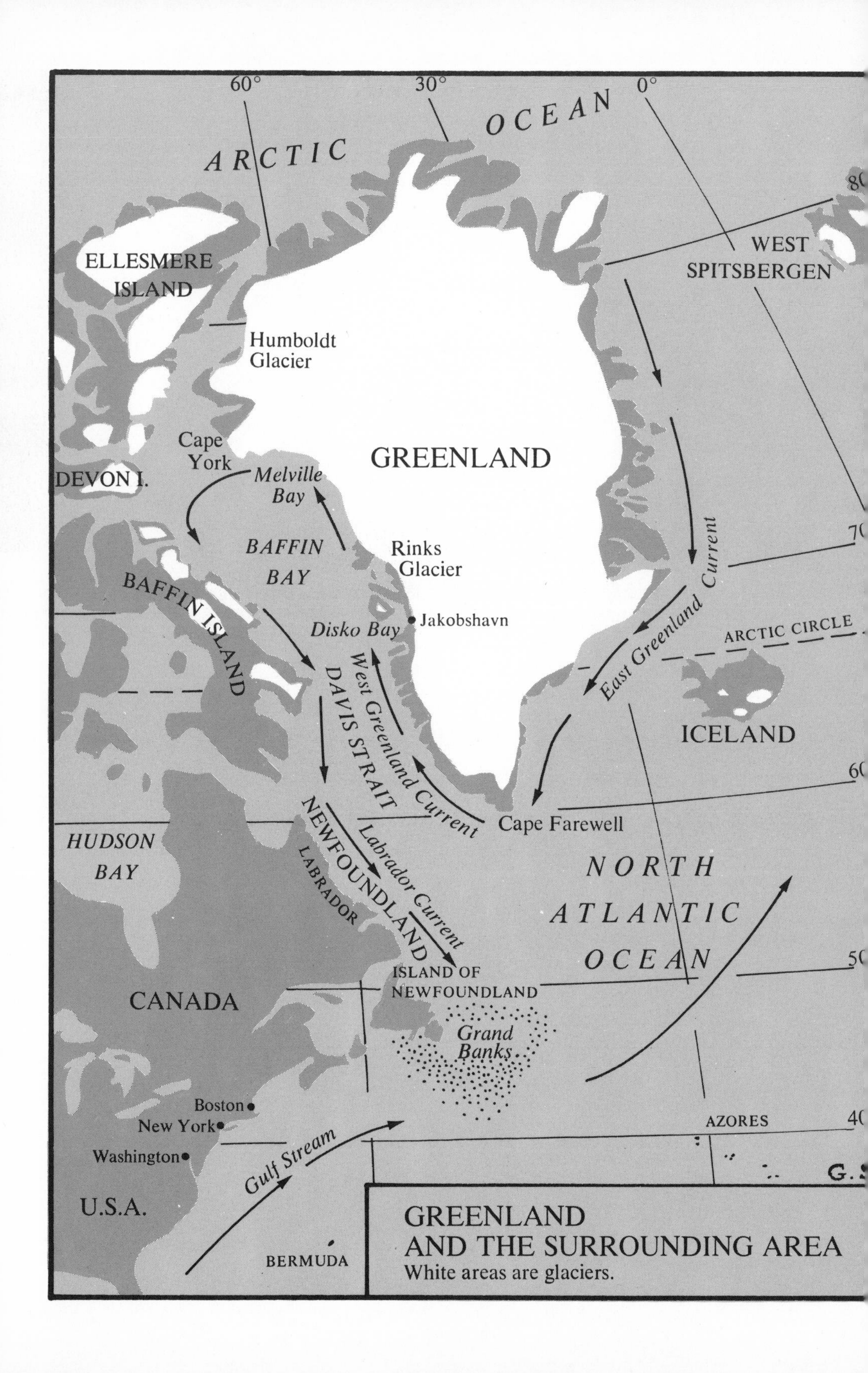

GREENLAND AND THE SURROUNDING AREA
White areas are glaciers.

Sprawling, thoroughly crevassed Jakobshavn Glacier on Greenland's west coast sends icebergs out on many sides.

the forty-eighth parallel (the latitude of the north tip of Maine) off Newfoundland. One extreme year (1939) 1,351 were sighted past that parallel, and another unusual year (1966) none was. The icebergs drifting south continue melting away, but each year during the warm

The luxury passenger liner, the *Titanic,* being escorted out of the harbor at Southampton, England, as it started its maiden voyage that ended in disaster on the Grand Banks off the island of Newfoundland.

season about thirty-five to forty of them reach latitude 42 degrees north (about the latitude of Boston) in the Grand Banks. That is approximately where the British luxury liner, the *Titanic,* sank. History will long remember that disaster.

The year was 1912, and the *Titanic* was a brand-new ship, the largest ship afloat, publicized as unsinkable. It was on its maiden voyage, one week out of Southampton, England. On that calm, moonless night of April 14, it was 300 miles southeast of Newfoundland approaching New York. Its captain had been warned several times of

This iceberg is apparently the one that the *Titanic* collided with, as it was the only one at the scene when the first rescue ships arrived.

ice in the waters, but he did not slow the ship. Probably he was trying to set a speed record across the Atlantic. And he may have been overconfident. His ship was the supreme ship on the seas and proclaimed indestructible. Its powerful engines pushed it on through the night. Then suddenly in the blackness an iceberg was sighted ahead! There was no time to avoid it. Only forty seconds later the ship rammed into it and was torn open from stem to stern. There were some survivors, but that night 1,517 people died in icy water in the worst iceberg-caused tragedy in maritime history.

A "bomb" of indelible, penetrating red dye is about to be dropped on an iceberg in Davis Strait off Baffin Island, so it can be identified as it travels. The berg's drift and rate of melting can then be measured.

The very next year, 1913, the International Ice Patrol was established. Its duty has been to locate dangerous icebergs and help ships avoid them. Using planes and ships, it locates possibly troublesome large bergs off Newfoundland and tracks them till they become too small to do damage. Spotters figure their speed and direction of travel, and this information is broadcast. Satellites are a help in scanning iceberg waters, and computers in predicting where individual bergs will drift and melt. Sometimes dye is dropped or shot onto a particular berg, so it can be identified by the splash of color.

The International Ice Patrol is financed by the main shipping nations of the world, but the United States was asked to manage it. Responsibility rests with the United States Coast Guard, which has carried out the work with vigilance and success. Since this ice patrol was begun,

Now the International Ice Patrol warns ships of dangerous icebergs in the Grand Banks area. These men in a lifeboat are crew members of the United States Coast Guard cutter to the right. It is alerting approaching ships by radio and light beacons of the presence of this iceberg, which is as long as an average city block.

there has been only one collision between a ship and an iceberg in the Grand Banks area, and not any ship or life has been lost (except for war years). However, there remain other unpatrolled areas of ice danger, where ships must exercise extreme caution and some have been lost due to iceberg collisions.

The iceberg-patrol season in the Grand Banks area is from mid-March to late July, the worst months being April and May. Sometimes it is necessary to begin patrolling in February or to continue as late as August. During the iceberg season, shipping lanes across the North Atlantic shift a little south to avoid the most dangerous zone.

Despite modern methods of detection and warning,

A huge iceberg off Jakobshavn, Greenland, frames a United States Coast Guard icebreaker.

A United States Coast Guard tug observes a picturesque but menacing berg in the North Atlantic.

cloud and fog continue to hide icebergs so they are still a danger. Before the use of radar, ships blew their whistles to locate bergs by echo. But neither of those methods is totally reliable. Ice is a poor reflector of radar waves, so the radar operator cannot always identify a berg even with a strong signal. Also, icebergs that are low escape detection. There is a further danger in the fact that below water level bergs may extend outward quite a distance, and if a ship hits an underwater ledge or point of ice, its hull can be ripped open.

Waters churning around this iceberg indicate its underwater base extends out well beyond this visible peak. Broken-off pieces float away.

One of the most reliable iceberg detectors is the lookout watching from the bow or, better yet, from a higher position on the ship. Ships have been saved when in the dark of night or dense fog a lookout felt a frigid wind, realized what its source must be, and immediately ordered the ship's engines reversed. Or, in olden days, ordered the oarsmen to change course.

The distance at which an iceberg can be sighted depends on several things: the height of the berg, the eleva-

A helicopter view past the 500-foot peak of an iceberg at Disko Bay, Greenland. A 269-foot icebreaker passes in the background.

tion of the observer, and the amount of visibility. If the lookout is at a height of seventy feet and the day is very clear, he can spot a large iceberg eighteen miles away. It will be quite brilliant then. With a low-lying haze around the horizon, lookouts have seen the tops of bergs nine to eleven miles away. In light fog or drizzling rain a berg is visible at a distance of up to three miles.

In dense fog a berg cannot be seen until it is treacherously close. With the sun shining through a heavy fog, a berg in the distance first appears as a luminous white form. In a sunless fog it appears a dark, menacing mass with a narrow black streak at the waterline. At night the waterline becomes a white streak. In a light, low fog an observer can see a berg sooner from aloft than from the deck, but in dense fog as a safety measure, there should also be a lookout in the ship's bow, peering straight ahead.

On a starry, clear night a lookout will not be able to see a berg farther than a quarter of a mile away, but if he has been alerted to its position he can with the aid of binoculars occasionally sight it at a distance of a mile. With a full moon he can locate a berg at a greater distance, possibly up to three miles.

Even when they are plainly visible, icebergs are not to be trusted. Sometimes part of one will suddenly break off, and this smaller chunk will sink and bob up and down violently, just as the mother iceberg did when it was born. Also, bergs become unbalanced as they melt, and when their center of gravity shifts, they turn over. Without warning their underwater spurs and ledges can tip up out of the water and anything in their way had better beware.

But in spite of their dangerous aspects, icebergs are sometimes a help to ships at sea. During storms ships

A plane of the International Ice Patrol keeps track of a dangerous berg.

often seek shelter in the lee of a big iceberg, where they are shielded from gale winds. Or a ship may be in waters that are thick with pack ice, and the wind may be blowing these pieces against the ship. If the ship can position itself on the leeward side of an iceberg, it can escape being banged by the ice floes because the berg, with most of its mass below water, is not moved as much by the wind. Or, a ship may be frozen tight in a surface layer of ice so it cannot move. Then a berg or several bergs driven by wind or currents plow through the frozen surface and create a lead, or open passage, freeing the ship.

Icebergs can be a life-saving source of fresh water for ships whose water supply is exhausted. A ship or lifeboat is brought alongside a safe-looking berg, and meltwater is collected from pools on its surface. This practice is recommended only in an emergency and was resorted to more in years past than now.

A tabular berg, 200 by 400 feet in size, with a pool of meltwater. Such pools of fresh water have often saved the lives of people at sea when they ran out of water.

Small icebergs including some bergy bits stranded at low tide in front of Taku Glacier near Juneau, Alaska.

Small bergs are not a serious menace. A modern ship can push them around if need be. There are special names for the very smallest bergs. Miniature ones, no more than about three feet across, are called *bergy bits.* Somewhat larger ones, which rise five feet or more above the water, are called *growlers.*

While icebergs are a threat to shipping in parts of the North Atlantic, the North Pacific is virtually iceberg-free. No iceberg-forming ice sheet like Greenland's reaches that ocean. No strong currents bring icebergs from sources far away. The Pacific is not open to the Arctic Ocean except at the narrow Bering Strait. Icebergs from Alaskan glaciers usually do not drift far from the coast, and those that do are relatively small.

Icebergs of the northern hemisphere have played an exciting role in the history of exploration and ocean travel. But the largest icebergs of all are at the opposite end of the world.

CHAPTER THREE

ANTARCTIC ICEBERGS

Antarctica is renowned for its huge, flat-topped icebergs, or ice islands, which originate as broken-off sections of ice shelves. It also has an abundance of the irregular bergs formed from crevassed, sloping glaciers. But ice islands are commonly associated with Antarctica because there they are exceptionally large and conspicuous. (After ice islands have aged and been altered by melting, wave action, capsizing, and splitting, they also assume more irregular shapes.)

There is a reason why ice shelves, and therefore ice islands, are larger and more common in the Antarctic than in the northern hemisphere. The south polar region is colder than the iceberg-producing areas of the North. Antarctica is situated right over the pole, while Greenland and other arctic islands are some distance from the pole, and Alaska's iceberg coast is in a temperate climate. Antarctica, still gripped in a true Ice Age, holds the world's largest, most powerful masses of ice. Its ice sheet

Glaciers draining off Antarctica here form an ice shelf—a flat, thick sheet of ice moving into and across the sea. This one is shattering, its cast-off pieces becoming icebergs as they drift away into the thinner pack ice.

is thicker and much broader than Greenland's. Many of its outpouring glaciers are still so cold and firm when they reach the sea that they do not break off there but continue pushing outward. They unite and advance over the sea as flat ice shelves, some of which are hundreds of feet thick and the largest in the world. In places the shelves rest on raised parts of the ocean floor or surround islands, and in that way they receive support. Elsewhere they just float stiffly upon the water.

The shelves are kept moving seaward by the force of the glaciers advancing off the land at the rear. The shelves are fed too by snow collecting on them, just as it does on land glaciers, and by water freezing on them from below. They move seaward, away from the shelter of bays and the coast, until they extend to where they are at the mercy of polar storms, smashing waves, and warmer temperatures, and there pieces of the shelf crack loose and drift off. Over a third of Antarctica's coast is fringed by such ice shelves, and they all produce flat-topped icebergs.

The Ross Ice Shelf is the king of ice shelves. Its area is as large as that of California. Its shape is triangular like the bay it occupies, and where it faces the Pacific Ocean it is 500 miles wide. It is 800 to 1,000 feet thick. Most of this thickness, of course, is below water, but about 150 feet of it is a sheer cliff rising above the waves, taking the

Wind blows snow off the outer edge of the Ross Ice Shelf, Antarctica's largest ice shelf, whose area is as large as California's. The shelf moves seaward about 3 to 6 feet a day as land glaciers feed it from the rear. Its ice-cliff edge is as much as 150 feet above the water and extends hundreds of feet below. Sections of the edge break off, creating icebergs.

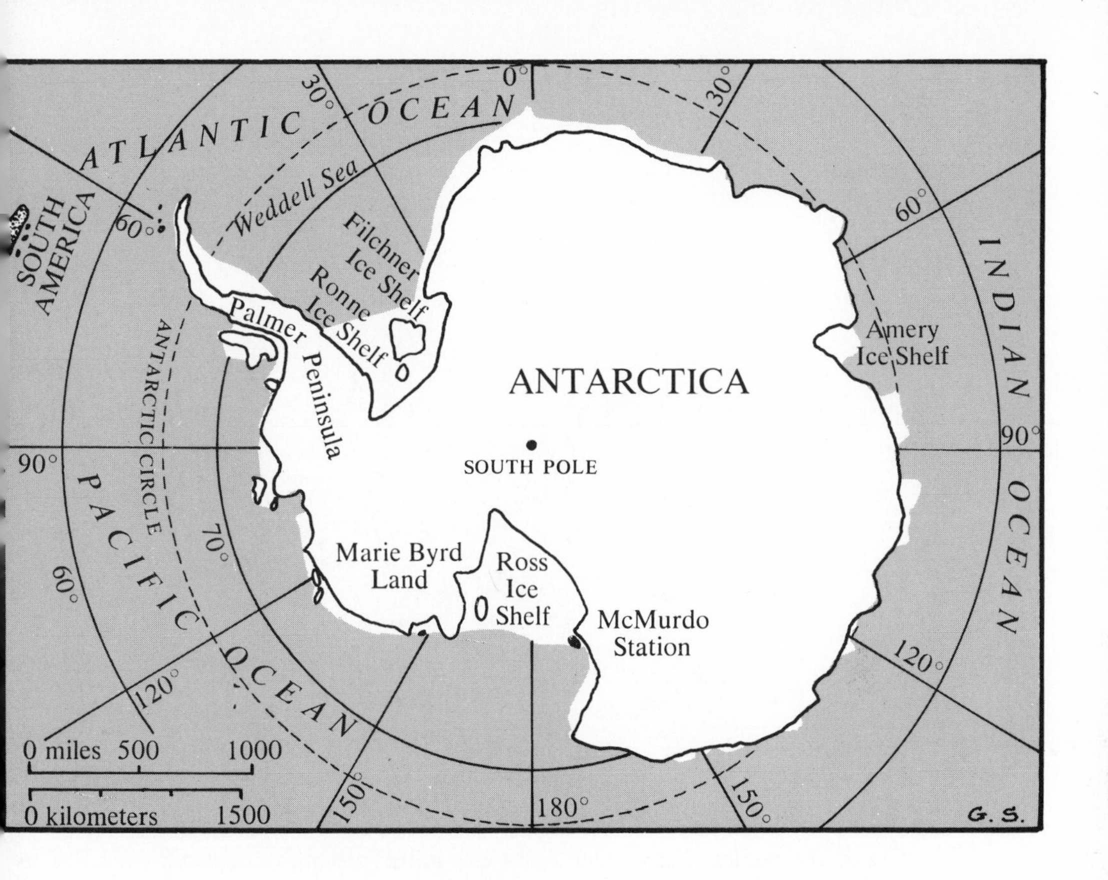

beating of the wildest storms on earth. The shelf's surface is smooth with some irregularities. The first antarctic station of the United States, Little America, was built on this shelf.

Other of the larger ice shelves along Antarctica's coast are the Edith Ronne and the Filchner ice shelves, which are joined and face Weddell Sea on the Atlantic Ocean side of the continent. They are also in a protected bay. Amery Ice Shelf faces the Indian Ocean. Some ice shelves have not yet been named.

A United States Navy ship moves past a tabular iceberg in the Weddell Sea of Antarctica. Many such bergs are miles in length and have been mistaken for land.

Antarctica produces supersized icebergs, some having lengths of 50 to 100 miles. Some contain more ice than does the world's largest valley glacier. One had an area twice that of Connecticut. Another was still larger, with dimensions of 60 by 208 miles. Such mammoth, flat icebergs are often mistaken for land.

These immense bergs take years to melt, especially if they remain in cold ocean currents that circle Antarctica. They may float around there for ten or more years.

A large iceberg in the water near Palmer Peninsula, Antarctica. Waves have been eroding it at water level.

Waters surrounding Antarctica are icy cold not only because of the cold atmosphere but because they are continuously chilled by the ice pouring into them from the continent, which may be thought of as a huge refrigerator spilling out gigantic ice cubes.

Icebergs freeze tight in those frigid waters during the long, sunless winter, and they move again as leads open in the summer. In time, they break into smaller bergs and are melted by waves and currents, at water level and below, and by moderately warm air masses that invade the region in summer. So they gradually become smaller and disappear.

Some are swept along by ocean currents flowing to lower latitudes (counterparts of the Labrador Current of the North). In warmer waters melting occurs faster. Many bergs drift into the Falkland Current coming north from South America's Cape Horn, and are carried into the South Atlantic Ocean. Some reach as far north as the latitude of Buenos Aires. No ice patrol checks their courses there as in the North Atlantic, because fewer ships travel in those seas, fog is not as common, and there are not as many icebergs as in the Grand Banks area. Rarely do icebergs move far enough north to enter shipping lanes in the Indian or South Pacific ocean, although they have been seen a hundred miles south of Australia.

The fiorded coast of southern Chile is the only producer of icebergs, other than Antarctica, in the southern hemisphere. These icebergs, however, are relatively few and small and do not travel far.

Antarctica's threatening, impassable fleet of icebergs helped keep that continent unapproachable and unknown until just the last century. But, paradoxically, if it had not been for the icebergs, Antarctica would have remained undiscovered still longer. The icebergs straying into

These icebergs with perpendicular sides and relatively flat tops have calved from an ice shelf along Marie Byrd Land, Antarctica.

warmer waters were what convinced explorers that there had to be land to the south, for icebergs come from glaciers and where there are glaciers there is land. But the way was blocked.

During the era of exploration and discovery, ships could not get near Antarctica. Terrible storms and ice-clogged waters made that the most hostile place on earth and prevented fragile sailing vessels with wooden hulls, wooden masts, and canvas sails from coming within sight of it. No airplanes or satellites existed then to fly over and see what was there.

Stronger ships were built, and they probed farther south. However, the farther they ventured the larger and more numerous the icebergs were. This discovery was discouraging but strengthened the conviction that land must be beyond, and the perilous search continued.

Finally, in 1820, the coast of the White Continent was sighted, and the next year it first felt the foot of a human being. But only in the last few decades has the iceberg blockade been pierced sufficiently so that the continent can be approached by ship with some degree of safety. And even now that continent and its ocean moat remain a realm of danger and mystery. In olden days, when ships were easily crushed by the ice, and when there were no radios and no rescue planes, only the bravest, hardiest adventurers dared think of going on an expedition to Antarctica, to the "end of the world."

The British poet Samuel Coleridge made an imaginary visit in 1798. His famous poem "The Rime of the Ancient Mariner" describes a voyage into the sinister waters south of South America on the fringe of iceberg territory. That poem is the one that contains the well-known lines: "Water, water, everywhere, nor any drop to drink." In the same meter he wrote:

> And now there came both mist and snow,
> And it grew wondrous cold:
> And ice, mast-high, came floating by,
> As green as emerald.

The American admiral Richard E. Byrd, in his book *Alone,* tells of the adventure in which he almost lost his life, when he wintered alone in the camp Little America. He describes his approach by ship through iceberg waters in 1934 this way:

> I don't think that any of us will ever forget what it was like in the Devil's Graveyard: the sunless corridors of waste waters; the fog that sometimes thinned but never lifted; the crash of the gales, and occasionally over that uproar the heavier sound of bergs capsizing in the storm; and everywhere those stricken fleets of ice, bigger by far than all the navies in the world, wandering hopelessly through a smoking gloom. Through this ambush the ship groped and side-stepped like a lost creature, harried by enemies

The National Science Foundation's research vessel *Hero* in waters off Palmer Peninsula, Antarctica.

her lookouts rarely saw full view, but only as dark and monstrous shadows sliding through the fog.

To ships approaching by sea, even in the modern era, those icebergs present a most imposing picture. In his book *Antarctica,* geologist Frank Debenham describes his first sight of the icebergs in these words:

> They are met long before one reaches the zone of the pack-ice and whether first seen on a clear sunny day or through the loom of fog the same idea of might

The *Hero* picks its way through icebergs off Palmer Peninsula, Antarctica, where many glaciers reach the sea.

and majesty spring to the mind. To that one may add the notion of purity since the upper parts are always spotless white and usually severely rectangular. One can hardly help investing such monsters with some kind of personality.

And he further says of an iceberg:

The play of sunshine on its towering sides, the blues and emerald hues of its cracks and crevasses defy description just as does the sense of solidity

> when the great waves of those seas break against the cliff without causing a shiver and throw their spray to the full height of the ice.

To him the Antarctic icebergs are "less beautiful perhaps than Arctic icebergs but more majestic."

Both types have their own kind of grandeur.

CHAPTER FOUR

ICEBERGS TODAY

The image of icebergs is changing. We still think of them as frightening—stark giants superior to us. But over the years we have begun to face up to them. We are far better equipped to do so now. We have more powerful steel ships with reenforced hulls. We have modern navigation and communication instruments. With airplanes scouting from the skies, ships can maneuver among the bergs, though still with great peril. Icebreakers now can make their way along the northernmost islands of Canada from Greenland to the oil fields of northern Alaska. Cargo

Ice Patrol men climbed aboard this iceberg off Newfoundland with spiked shoes, ice axes, and lifelines—like mountain climbers—to conduct demolition tests. They drilled holes in the berg and put explosives in them.

ships regularly push through the bergy waters of Antarctica and dock at its coast near the Ross Ice Shelf. Even tourists go into Antarctica every summer. If we cannot get through the iceberg-controlled waters, we fly over them.

As the cold regions are being opened, the menace of icebergs is being met. We have learned to go around, among, and over them. Next we are trying to learn how to destroy those we want to get rid of. Direct attacks on icebergs have taught us just how tough and indestructible they really are. So far the only power that can destroy

The explosions that followed made beautiful fireworks but did little damage to the 75- by 100-foot iceberg.

one is the warmth from the sun, either in the atmosphere or in the water, and it works slowly. Some impatient experimenters have tried to destroy icebergs faster but with poor results.

They have attempted to break up bergs by bombing them, and that method has proved impractical and overly expensive. When bombs are dropped on icebergs, smoke and a few holes are made or the top is blasted off; if the center of gravity shifts, the berg adjusts, and more ice from below water level rises to replace that which is lost. The berg remains. Underwater torpedoes have had as

A cloud of black smoke marks a direct bomb hit on this iceberg, which weighs about a million tons and measures 150 feet high and 300 feet wide above water.

little effect. And explosions of underwater mines caused showers of ice but little reduction in the berg's size.

Still, these unsatisfactory results will not stop the experimentation. One way to do away with unwanted icebergs might be to increase the sun's warming effect on them. Pure-white bergs have the highest reflective power

This picture taken after the blast shows that the berg suffered little. The black dust left on it will probably hasten its destruction (by absorbing solar heat) more than the bombing did. During iceberg-destruction tests by the International Ice Patrol sixteen icebergs were bombed, but the damage to them was rated as "insignificant," proving the strength of icebergs.

of any natural thing on earth. They reflect a maximum amount of the energy received by the sun, well over 90 per cent. Therefore, very little is absorbed by the iceberg and available for melting. It is correctly reasoned that if a berg is dusted with dark powder, it will absorb more of the sun's rays and so melt sooner.

Some mountain glaciers near arid regions are purposely dusted with dark substances to make them melt faster and release more water to streams flowing from them. Also, frozen waterways in certain cold regions are dusted with dark powder to make them thaw more quickly in the spring. This powder may be soil, coal dust, or any ground-up dark material.

Half an iceberg was covered with black dust in a test to see how the dust would speed up the ice's melting. The job finished, the men hurry to leave the berg, carrying an extra lifeboat with them as a precaution. The island of Newfoundland is in the background.

Scientists are considering the possibility of placing huge mirrors in space and focusing intensified rays of the sun on selected cold areas—which could include iceberg areas —to melt away ice. The big question is whether doing so would improve conditions as hoped, for when we tamper with the environment, we may bring havoc where we intended to bring improvement.

While some people are trying to do away with icebergs, others are finding that they have certain desirable characteristics. Iceberg ice is becoming a sought-after commodity. It has special properties not found in ordinary ice that is made artificially in a freezer or cut from frozen lakes (the way ice is obtained where mechanical refrigeration is not available).

For one thing, iceberg ice is much colder. (Ice can be considerably colder than 32 degrees Fahrenheit, which is just its highest possible temperature.) Therefore, it makes beverages much colder and keeps them colder longer. In addition, this ice "sings" when put into a liquid and makes a beverage especially enjoyable and delicious. As the snowflakes and tiny ice grains that originally formed the ice were covered by more snow, some of the air about them in little openings was trapped and could not escape. Slowly the snow and ice were compressed into solid ice, and the bubbles of air were compressed too. Ice in the depths of a glacier is under tremendous pressure. A 1,000-foot thickness of glacier ice exerts a pressure of about twenty-eight tons per square foot, and some of this "singing" ice comes from even greater depths. When that compressed ice is placed in a beverage and starts to melt, the long-entrapped bubbles begin expanding, and as they are released they make popping, crackling, fizzing sounds. Some of these escaping bubbles of air were trapped in prehistoric times and ancient periods of history. One thinks of this as one looks at the glassy ice and listens to its tantalizing sounds.

Because iceberg ice is intriguing and unusual, people will travel out of their way to get it, and it is shipped long distances to be sold. It is now one of Greenland's exports. At Jakobshavn small bergs are towed to harbor, and there chopped by machine into ice cubes, which are sent to eager buyers in cities around the world.

Many lodges and resorts for winter and summer vacationers are built near glaciers. The operators of these places will often go to the nearby glaciers, and to small

Danger-defying men climb onto a grounded iceberg, perhaps to get ice. They will not stay long, for behind them the crevassed front of Mendenhall Glacier (near Juneau, Alaska) threatens to send new icebergs plunging into its lake.

icebergs in lakes alongside them, to chop off ice. They use it in beverages served to their guests as an attraction.

Iceberg ice is sought after also for general refrigeration purposes. Ships will approach a fairly safe small berg and chop off parts to keep their perishables cold. Tourists who can find an iceberg stranded on a beach, or can boat up to one, will chop off ice to use in their camper iceboxes or to pack around freshly caught fish.

So icebergs are not merely picturesque features of the environment. They are economic commodities and can be put to various uses. The following contemplated use is quite extraordinary, but it never materialized.

In 1942, when the German Army was still powerful in Europe and North Africa, the German submarine fleet was a serious threat to the Allies' ships. One of the most difficult of the Allies' problems was their lack of air power in battle areas far from their home bases. Airplanes of that time had a shorter flying range than today's planes and could not fly far enough to assist the Allied forces where they badly needed them. If the Allies tried an invasion on a distant shore they would be sorely handicapped by lack of air support until they had established their own airfields there. The planes operating from aircraft carriers were slower and less well-armed than land-based planes. Therefore, extra-large aircraft carriers with long runways were needed that could serve as airports at sea for regular land-based planes, which then could operate thousands of miles from home. But designing and constructing these supercarriers would take years—too long to wait. Besides, the required metals and other strategic materials were in short supply or unavailable.

In the fall of that year someone proposed an astonishing plan to the Chief of Combined Operations—to use icebergs as aircraft carriers. The plan was to smoothe their tops to make runways and to hollow them out to make hangars. Winston Churchill issued a directive that research on this secret project should go ahead full speed and have the highest priority. The bergships, as they were called, would be given power to steer and move slowly, and they would be protected against melting until they had fulfilled their assignment. Military strategists were excited about this daring operation for several reasons. These bergships would be inexpensive, they would use little critically scarce material, and unlike steel carriers they could not be sunk.

However, the war turned in the Allies' favor before the bergships came to be built. Apparently this proposal marks the first time serious thought was given to using whole icebergs for a practical purpose. Now there is talk of using icebergs in a still more fantastic way.

CHAPTER FIVE

ICEBERGS TOMORROW

The world is suffering from a thirst for fresh water. Most of its water–over 97 per cent–is salty ocean, unusable for agriculture, industry, and home and personal needs. So less than 3 per cent of our water is fresh, or nonsalty. Most of that amount–over 2 per cent–is in glaciers. All the water in lakes, rivers, the ground, and the atmosphere together makes up a tiny fraction of the world's fresh water. We do not think of glaciers as great freshwater reservoirs, but they are. And their offspring, icebergs, drift away to nowhere, letting their precious substance trickle unused into the salty seas.

Over the years imaginative minds have dreamed of bringing icebergs to areas that urgently need water and letting them melt there to deliver their valuable cargo. Knowing that icebergs occasionally drift into the subtropics on their own, it was reasoned they could retain more of their bulk after traveling such distances if they were steered there directly instead of taking a wandering course.

Transport icebergs to given locations in warm regions? And get them there before they melted to worthless size? The idea seemed like science fiction. Yet—considering the facts that icebergs are much colder than ordinary ice, they melt more slowly, and they travel with 85 to 90 per cent of their mass underwater—there might be a chance.

Early trials were encouraging. Between 1890 and 1900 small icebergs from southern Chile's fiords at latitude 45 degrees south were towed by ship to Valparaiso on the coast of central Chile and even as far as Callao, Peru, at about latitude 12 degrees south. But bigger icebergs were needed to make longer trips. Irregularly shaped ones would be unwieldy and hard to control. Those being eyed were the huge, stable, flat-topped icebergs from Antarctica.

An excellent place to market an iceberg, it was felt,

would be arid southern California because of its phenomenal population increase, expanding areas of irrigated cropland, and industrial growth. But the idea of bringing an iceberg to southern California, or anywhere, did not seem practical.

Wouldn't the iceberg melt before reaching its destination? Were there any ships powerful enough to propel and steer it? How could it be brought to shore? Since most of it is underwater, it would ground on the ocean floor before it got close. Somehow it would have to be moored offshore, and the water brought in to land. With all those difficulties the scheme seemed preposterous.

It would be a tremendous job to tow ice islands such as this one—reservoirs of fresh water—to arid coasts thousands of miles away, but engineers say it now can be done.

However, the world's population kept multiplying, and the need for fresh, clean water kept increasing. Regular water supplies became polluted. Desalting of seawater proved expensive and inefficient, and desalted water is not totally pure and not as desirable as really fresh water.

And so the dreamers looked again at icebergs. The water they contain is as clean as any that nature produces. Bergs that come from polar glaciers are especially pure. Their ice formed from snow that fell long before the industrial and scientific revolutions began filling the air with harmful pollutants, so the inner parts are immaculate and uncontaminated. Even now little harmful fall-out reaches remote glaciers like those of Antarctica. Tests made of water taken from ice in Greenland and Antarctica showed it to be nearly as pure as the purest laboratory water. A bacterial count at the South Pole showed only one bacterium per pint of snow.

About twenty years ago American researchers started to make estimates and plans. One study showed that if six ocean tugboats towed a large iceberg from antarctic waters to Los Angeles, the cost would be a million dollars. But the meltwater would be worth a hundred times that much. The trip would take about a year, and during that time the berg would have lost about half its volume. It could be grounded offshore, and a floating dam, extending below the surface, could be put around it. The fresh meltwater would stay on top of the heavier salt water and could be pumped to the mainland.

Russian geographers in 1969 described how icebergs could be used as water supplies for coastal cities. They figured an iceberg a mile long, a third of a mile wide, and about 500 feet thick contains about 150 million tons of water. That amount is enough to supply each person in a city of 8,000,000 with over 1,000 quarts a day for a month. They felt that although towing icebergs and tapping their water would present major technical problems, the feat theoretically could be accomplished.

Recently Chile announced a plan to bring icebergs from Antarctica to provide fresh water for cities in the Atacama Desert. Its mining port, Antofagasta, just south of the Tropic of Capricorn, is to be the first to receive this help. Its water now is piped from the Andes, but the supply is too small and not dependable.

In 1973, the United States Army Cold Regions Research and Engineering Laboratory raised the hopes of potential iceberg importers. It stated that it has become possible for supertugboats to deliver an antarctic iceberg to Australia or southern South America. The berg could be large enough to irrigate more than 6,000 square miles of land and would be worth over one billion dollars.

Iceberg towing is viewed with increasing optimism lately, because stronger tugs that can better withstand the pounding of ice and storms are being built. And the more powerful engines, developed for icebreakers and giant aircraft carriers, can be used. Also, Antarctica now has permanent stations; its seas are crossed by many ships and planes; it is not as forbidding as it used to be.

A close look at an atlas shows that certain geographic conditions are favorable for iceberg towing. Many deserts border on oceans. And many of these coastal deserts have

Icebergs are not as invincible as they used to be. Three United States Navy icebreakers push a flat-topped one out of the shipping lane leading to McMurdo Station, Antarctica. It measures 200 by 800 feet and rises 80 feet above the ocean surface. Around it are ice floes, the result of the breaking up of the surface sheet of sea ice.

cold offshore currents, which come from a polar source. (In some cases these cold currents are the cause, or part of the cause, for deserts existing there; they cool the base of air masses crossing them, making them stable and less able to rise and create precipitation.) Icebergs brought from Antarctica could travel much or all of the way in a cold current to reach desert coasts of western Australia, southwestern Africa, and South America. Those being taken to California could travel in a cold current as far north as the equator. The currents would provide much of the energy to transport the bergs and would cause less melting than warmer waters would. The prevailing winds move in the same direction as the currents along some of the routes and would furnish additional driving power.

The cold Peru Current flows from the antarctic to the equator along the western shore of South America. It is the cause of the coastal desert of northern Chile and Peru, including the Atacama Desert, which is the driest in the world. Along the east side of the South Atlantic the circulation is similar. There the cold Benguela Current flows north along the desert coast of southwestern Africa. Cold currents also flow to the arid coasts of western and southern Australia and to dry Patagonia of southern Argentina.

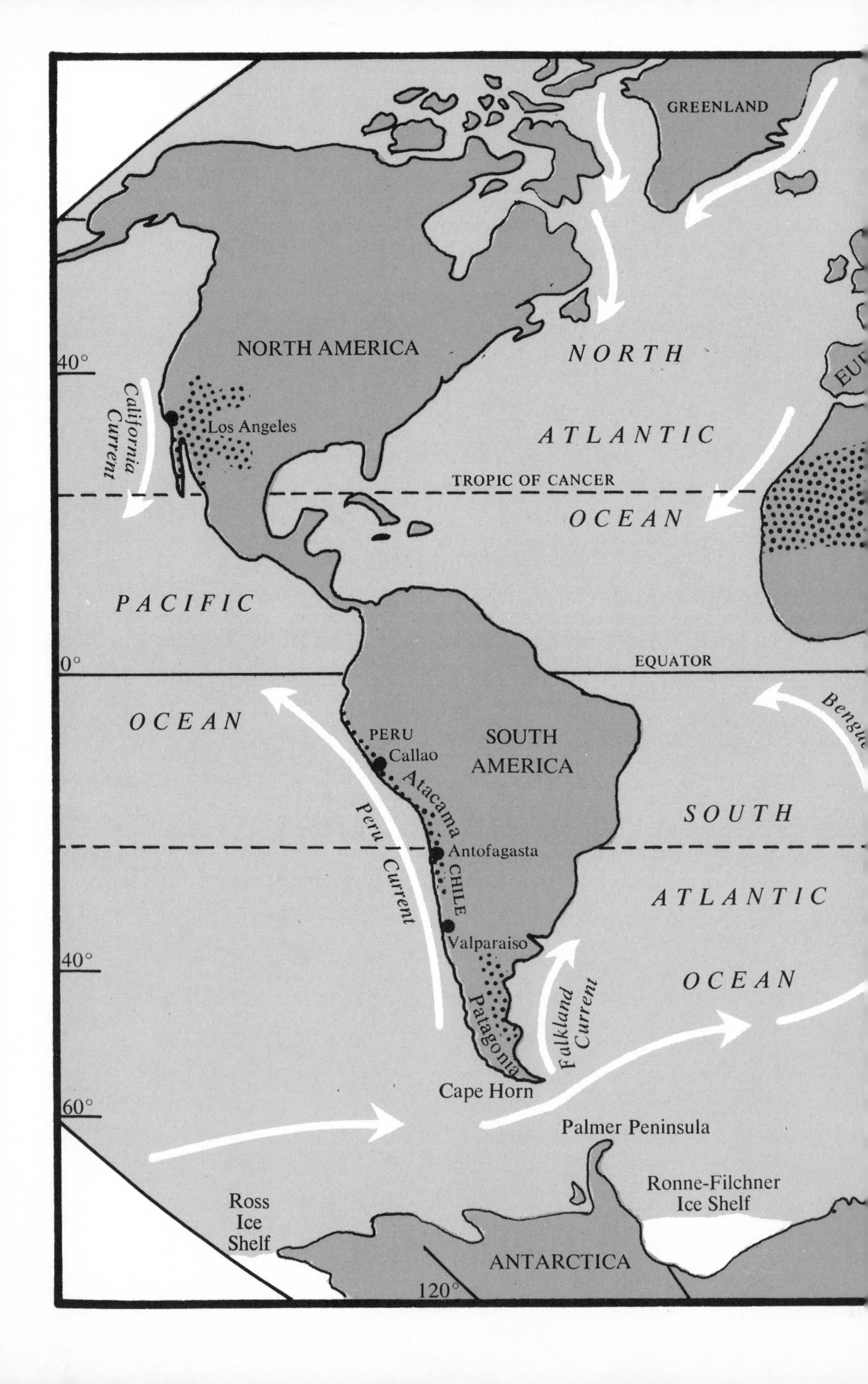
GREENLAND
NORTH AMERICA
NORTH
ATLANTIC
OCEAN
40°
California Current
Los Angeles
TROPIC OF CANCER
PACIFIC
0°
EQUATOR
OCEAN
PERU
Callao
SOUTH AMERICA
Atacama
Peru Current
SOUTH
Antofagasta
CHILE
ATLANTIC
Valparaiso
40°
OCEAN
Patagonia
Falkland Current
Cape Horn
60°
Palmer Peninsula
Ross Ice Shelf
Ronne-Filchner Ice Shelf
ANTARCTICA
120°

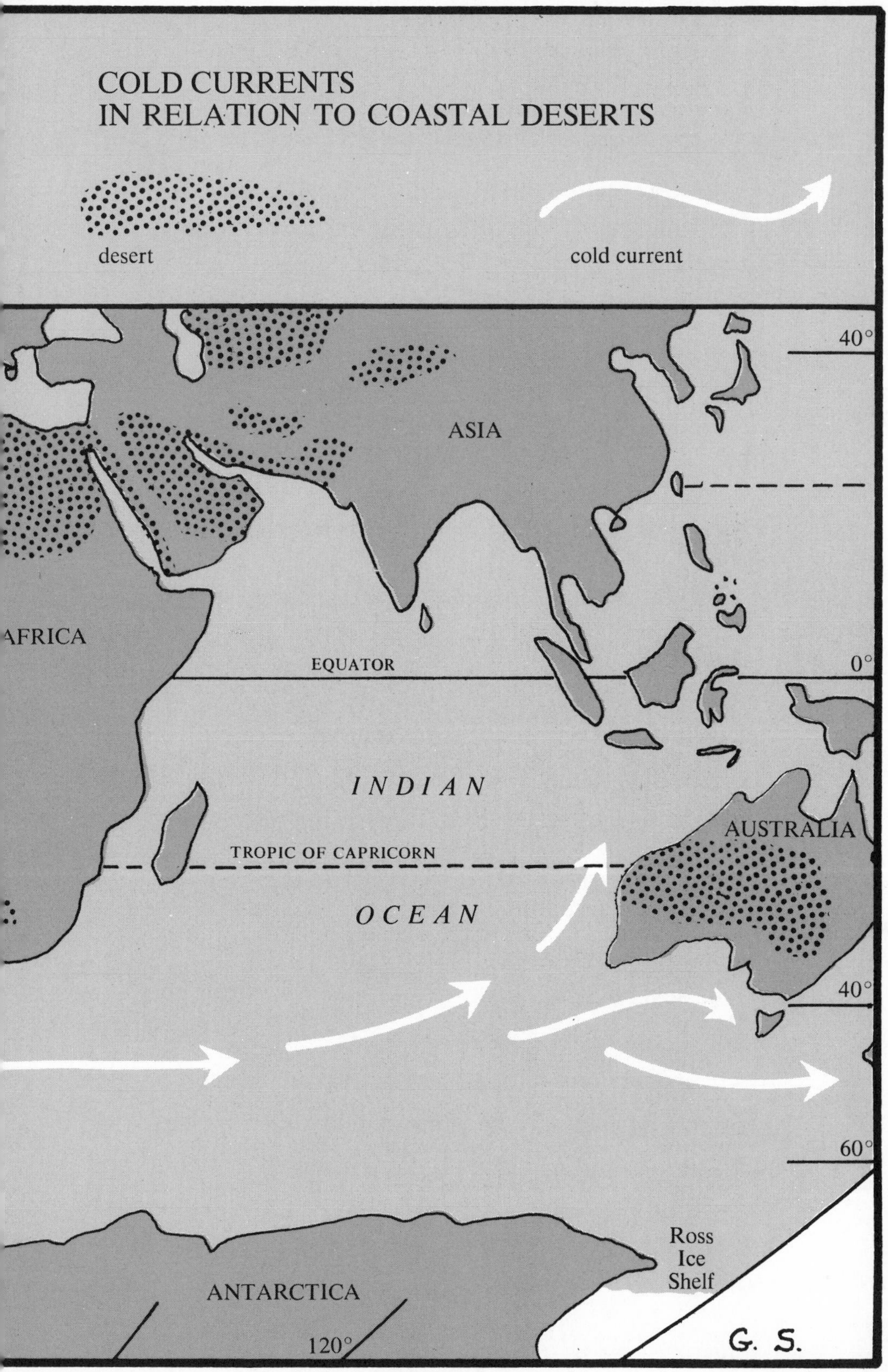

COLD CURRENTS
IN RELATION TO COASTAL DESERTS

desert

cold current

ASIA

AFRICA

EQUATOR

INDIAN

OCEAN

TROPIC OF CAPRICORN

AUSTRALIA

40°

0°

40°

60°

Ross
Ice
Shelf

ANTARCTICA

120°

G. S.

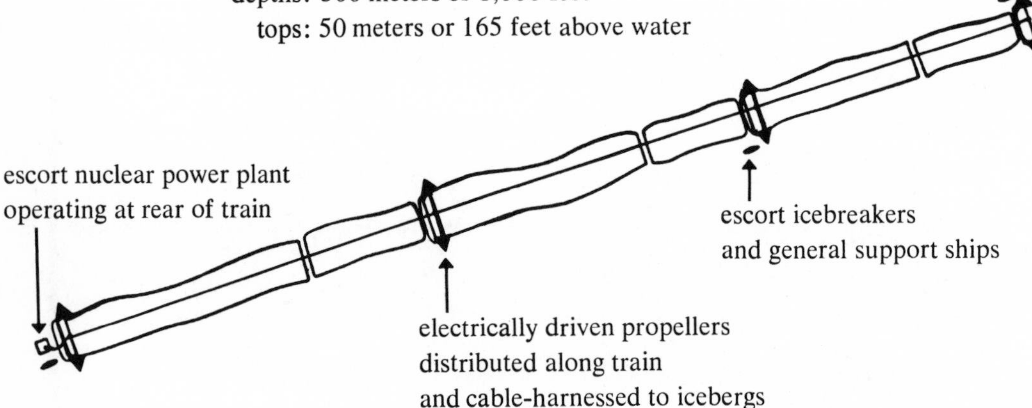

A drawing of how a train of icebergs might be assembled for its trip into warm regions.

One plan for towing icebergs from Antarctica is to link them together like a string of barges. The icebergs selected would come not from the outer edge of the iceberg zone but from near the coast of Antarctica, because there the bergs are larger and more numerous. The best choice of bergs with the wanted size and shape can be made right where they break off the ice shelves. Ross Ice Shelf is the most accessible.

The several bergs making up a train must all be solid, without cracks or other weaknesses, and must have the right proportions to "sail" well without danger of capsizing. They should be ship-shaped—narrow and long with no protruding parts to cause excessive friction and drag. A train could be 1,000 to 4,000 feet wide and about 12 miles long. The bergs would be a few hundred feet thick. They would be selected during the daylight season when the ocean was open, from January to March. Already icebreakers nudge some bergs around, and it is said a small single one could be picked up there any time of the year.

The job of selecting icebergs is easier now that satellites are photographing the area regularly. By referring to these pictures, large icebergs can be located more quickly and cheaply than by scouting with planes or ships. Clusters of proper-sized icebergs would be looked for, so

An iceberg of the shape that might be suitable for towing drifts in the Ross Sea near the Ross Ice Shelf, illuminated by the low sun of the Antarctic summer.

a train could be assembled in as small an area as possible. Final selection would be made after close on-the-spot inspection.

Then the bergs would be insulated and linked together. They would be insulated not just on top but underneath too. How this would be done is another major problem, but one suggested method is this. Each iceberg would be wrapped in plastic, double-layered sheeting with quilted pockets in it. The pockets would trap meltwater and so serve as insulation between the berg and seawater. This sheeting, coming from large rolls, would be wrapped under and over the berg and fastened on top by means of a

---

Satellite pictures such as this one can be used to locate icebergs that could be towed. This view from space shows a cluster of bergs partly locked in sea ice at the east end of the Ross Sea off Cape Colbeck, Antarctica. The sea is black. Clouds cover the left side. The scene is about 185 kilometers (115 miles) wide. Icebergs can be distinguished from ice floes in several ways. The higher icebergs stand out in relief against the water-level sea ice, so under magnification they are easily recognized and measured, and their shadows can be seen on surrounding sea ice though not on open water. The edges of tabular icebergs are usually crisp while those of ice floes, are usually eroded. Also, over a period of time, icebergs, which are steered by deep currents, can be seen to move differently from ice floes, which are moved by wind mainly, and often bergs will sweep aside sea ice and leave open water behind them. This is happening at the center of the picture where two elongated bergs making a right angle are moving south, toward the bottom of the picture, sweeping sea ice before them. Bergs suitable for towing would be among the smaller ones visible here.

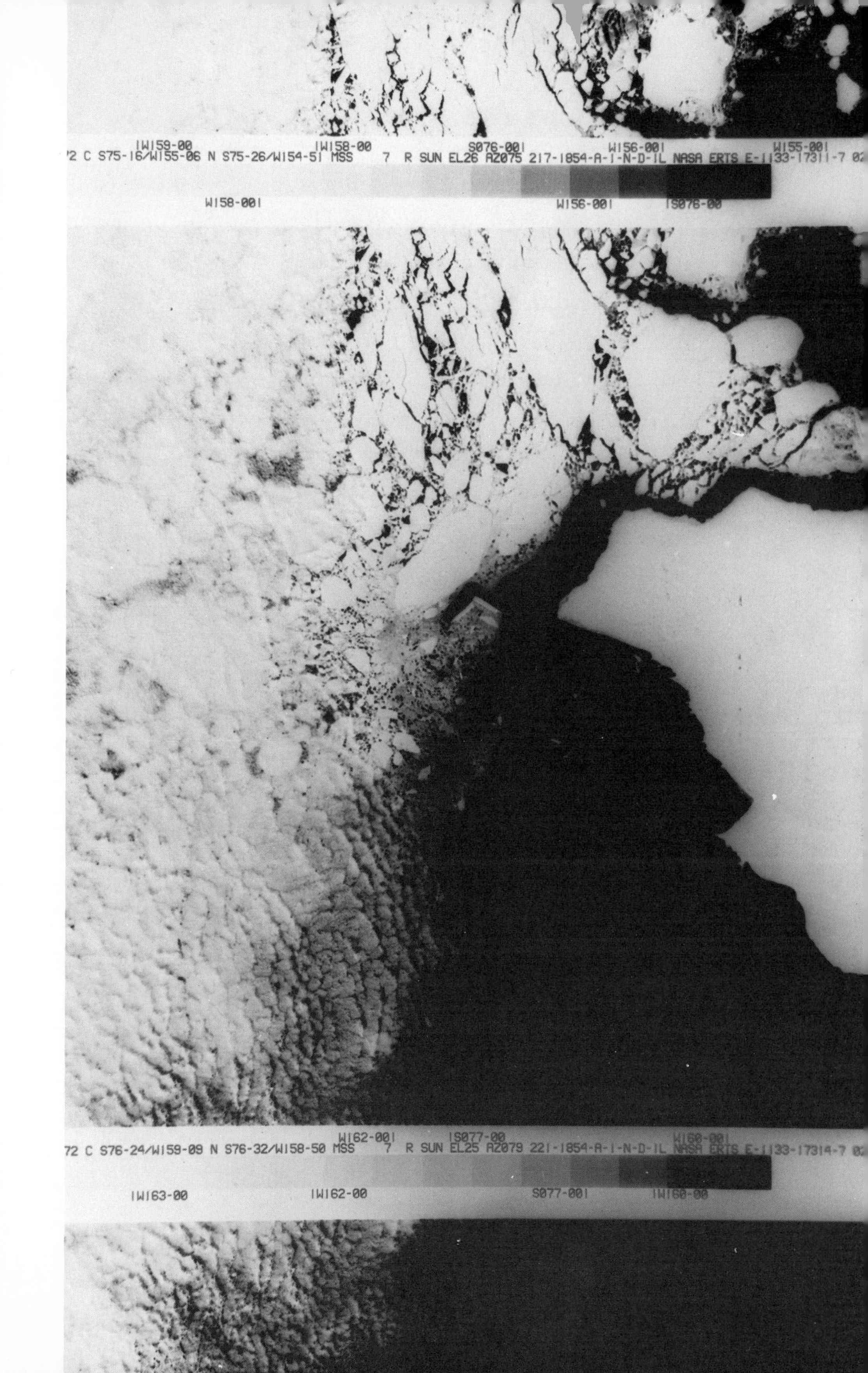

W159-00 W158-00 S076-00 W156-00 W155-00
72 C S75-16/W155-06 N S75-26/W154-51 MSS 7 R SUN EL26 AZ075 217-1854-A-1-N-D-1L NASA ERTS E-1133-17311-7
W158-00 W156-00 S076-00
W162-00 S077-00 W160-00
72 C S76-24/W159-09 N S76-32/W158-50 MSS 7 R SUN EL25 AZ079 221-1854-A-1-N-D-1L NASA ERTS E-1133-17314-7
W163-00 W162-00 S077-00 W160-00

weighted cable passing under the berg between two tugs. The cable would loop around the berg and pull the sheeting from the bottom side to the top. The sheeting need not be watertight. It would be put on starting from the back of the berg, and the next layers would overlap like fish scales toward the front. In this way the iceberg would be streamlined to the water flow.

An estimated 100 to 300 hours would be needed for wrapping a large berg. The first bergs to be wrapped would be those to which propellers would be attached. Power for the propellers and electric motors would be drawn from escort ships. The iceberg train would then be harnessed together with encircling cables and nets, and other bergs would be wrapped while the convoy was moving.

With that insulation, it is said, melting might be limited to less than ten per cent during a trip to Los Angeles, which would take about a year. Traveling along with the slowly moving iceberg train would be tugs, launches, helicopters, and a work crew. Because the bergs are large and free-floating, they would ride out the highest waves without rocking.

Let us assume an iceberg train reaches its destination. What is done with it? The train could be left together and anchored offshore, where it could serve as a long breakwater. Or it could be taken apart. Even then the large bergs would have to remain some distance offshore. Several ways are suggested for catching the meltwater and bringing it to land. The floating dam, mentioned before, is one. Or the ice could be quarried or broken into chunks, and the pieces brought in, perhaps on conveyor belts. Or water might be drawn directly off the ice as it melted. It could be made to collect in natural or dug-out hollows, or tunnels and gullies could be carved in the ice to channel runoff to collection points. When the bergs become small, they could be shoved onto a shoal or shore or into a diked-off bay or lagoon.

Offshore icebergs would serve as better reservoirs in some respects than lakes and ponds. They do not cover land that is needed for other purposes. They do not fill up with silt. They would lose less moisture by evaporation and, in fact, would acquire additional moisture by gathering condensation from the atmosphere because of their coldness. Although they could not serve as recreation spots as lakes do, they would be a great tourist attraction in a harbor. Also, iceberg ice could be sold and used in many interesting ways.

Imported icebergs are bound to affect the environment. They may create cool breezes and fog, and they may lower the local air temperature a little. The water about them would cool, and marine plants and animals would be affected by the temperature and salinity changes. The cooling could be advantageous or not. Cold water in itself is not harmful to sea life in general. In fact, animal life is usually more abundant in cold waters than in warm. What biological readjustment might occur in the water would have to be studied in each situation. While the presence of one iceberg might have only a passing effect, more permanent changes would take place if icebergs were present continuously.

Iceberg importing would have international effects too. If selling icebergs proves to be possible and profitable, competition will begin. Icebergs in international waters are available to anyone. Competitors will stake claim to certain bergs, perhaps a year or more in advance of using them, and station personnel "on board" to assure ownership. If fresh, pure water continues to become ever more scarce, there will come a time, some predict, when most of the desired icebergs will be snatched away from their home waters as soon as they are born.

Already we hear planners speak of "harvesting" icebergs, and we realize those floating fortresses are being given the status of usable resources, like trees or iron ore. How much longer can they remain independent sovereigns of the sea?

EPILOGUE

Icebergs are one of the last natural things to resist the assault of civilization. But civilization is now closing in on them.

Already they are blasted and bombed in attempts to make them fall apart. They are dusted with dirt to make them melt faster. Their virgin-white surfaces are splotched with dye so they can be identified and tracked. They are chipped for party ice. Pieces of them are used for refrigeration. Tourists come in ships to glacier fronts to see calving icebergs splash into the water, and if none falls, a gun is shot or the ship's whistle is blown loudly to create vibrations so the ice will loosen and fall for the visitors. Soon, it seems, icebergs will be harnessed and taken like captives to foreign waters to melt.

Something of regret comes over us when we think how these mighty monarchs, invincible since the Ice Age, are about to lose their old power. The regret is the same that we feel as we see a proud wild animal caged, a stately old tree sawed down, or a champion athlete going down to defeat.

Just as snows that made today's icebergs were falling while prehistoric people were emerging from the Ice Age or laying the foundations of history, so snows now falling on glaciers may drift forth as icebergs thousands of years from now when this planet will be quite different from what it is now—better or worse, we cannot say. Perhaps by then earthlings will have spread to other planets. Maybe then icebergs will again have their waters to themselves. Or perhaps the climate will continue warming, naturally or because of human interference, and the glaciers will shrink back from the coasts and from the lake edges, and icebergs will be no more.

Whatever the future, let us appreciate our icebergs now, as they are today. While we may, let us still see them as the romantic marvels they are. Picture them as handsome ships that are dramatically launched, that take their course singly or in fleets, and like ancient high-masted galleons of the past, set sail from their home harbor, never to return.

Whatever exotic sights there be in the universe, surely none can surpass the mystic beauty and allure of icebergs asail upon the seas of Earth.

INDEX

**indicates illustration*

INDEX

GWEN SCHULTZ is an associate professor of geography at the University of Wisconsin, Madison, currently serving as a writer for the Wisconsin Geological and Natural History Survey.

Ice has been a favorite subject of hers. She has written two previous books about it: *Glaciers and the Ice Age* (Holt, Rinehart and Winston, 1963), and *Ice Age Lost* (Doubleday & Company, 1974). And she enjoys doing field work and traveling in areas of glaciers and icebergs. She also has published a story for beginning readers, *The Blue Valentine,* with William Morrow & Company, and other books as well. Among her writings are many professional and popular articles on geographical and other subjects.